Copyright © 2019 by R⸺

All rights reserved under the Pan-American and International Copyright Conventions. This book may not be reproduced, in whole or in part, in form or by any means electronic or mechanical, including photocopying, recording, or by any information storage and retrieval system now known or hereafter invented, without written permission from the publisher, The Kings' Press, LLC.

Published in the United States of America
First published as The Kings' Press, LLC paperback 2019

For information about permission to reproduce selections from this book, write to:

The Kings' Press, LLC
933 Louise Ave. Ste 422
Charlotte, NC 28204

OR

Email:
thekingspressllc@gmail.com

Subjects: LCHS:
Outer Space | K-12 Education | African American Authors | Science Education | NASA for Kids | Children Books

Editing done by Ron King Jr. in collaboration with Josiah King

Contact Email: thekingspressllc@gmail.com
ISBN: 978-1-7368791-1-5

TABLE OF CONTENTS

About the Author...3
Acknowledgements..5
Introduction..6
Galaxies..7
Planets...9
The Sun..11
Black Holes..14
Solar Flare..15
Space Debris..16
Meteors, Asteroids & Comets...18
Supernova..22
Solar Eclipse..23
Josiah's Quiz..24
Glossary...25
Special Projects...28
Tutorial Test Key..33

ABOUT THE AUTHOR

Josiah's journey for writing this book about space science was a blessing in disguise. in the fourth grade, Josiah attended a school where fourth graders did not learn science for the majority of his school year. Not learning science in any grade was unheard of for his mother and i. As his parents, we were upset to learn that a school decided not to teach science. We didn't want our son to become discouraged in his studies, so we did what parents naturally do. We advocated for him.

Josiah too, was very disappointed about not learning science, and did not want to go to school as a result. His mother encouraged him, and was diligent in making sure Josiah's teacher, principal, zone superintendent, district superintendent, and elected school board member were aware of the situation affecting our son's education. After many discussions not much was done to accommodate Josiah and his love for science. Our appeals seemed to fall on deaf ears.

Earlier in the school year, Josiah and i talked about his options as we normally do. Okay i admit, it was more like a lecture. i suggested during those conversations, that Josiah pursue science himself. i told him he possessed the most critical skills one could have. He did not realize he could use the powerful skill of reading.

in the words of Booker T. Washington, Josiah did not allow "his grievances [lacking science], to overshadow his possibilities [in learning it himself]." Soon, he began to realize he can absorb anything with the right training and no one, or school in this case, could stop him from learning. The rest, as we say in the United States, is history.

it was a school project assigned at the end of the year that served as a catalyst for this book. When he showed me what he was working on for his class presentation, i then suggested he publish his first book. His mother, brothers, and i, are extremely proud of this achievement. His creativity has been impressive to watch develop over the years. He looks forward to publishing many more in the future.

We hope you and your family enjoy reading and studying the outer space science he puts forth in his very first book.

OUTER SPACE & ME

JOSIAH KING

Front cover design by Briana Osaseri

ACKNOWLEDGEMENTS

i want to thank my mom for supporting me. She is always available for me and dedicating her time to speak up for me at school.

i also want to thank my dad for supporting me and dedicating his time to edit this book.

INTRODUCTION

This book is about things that interest me when reading about outer space. if i were to take a trip to outer space, these are things i would like to see for myself. Galaxies, black holes, our solar system, meteors, comets and asteroids all sound and look very cool to me. Although solar flares, a solar eclipse, supernovas and space debris can be scary and dangerous, i would still like to see outer space in action.

Space science is interesting to me because i get to learn and explore things that do not happen on earth. Astronomers & astronauts teach us a lot about what's going in outer space. What i like about outer space the most, are the things not normal to earth.

The weather conditions, and resources are really interesting too. i would like to learn more about life on other planets, and how scientists collect information from planets that are far away. Eventually, i would like to take a trip to the National Aeronautics Space Administration's (NASA) space centers. May be we can go together and explore outer space.

Thank you for your support and i hope you enjoy!

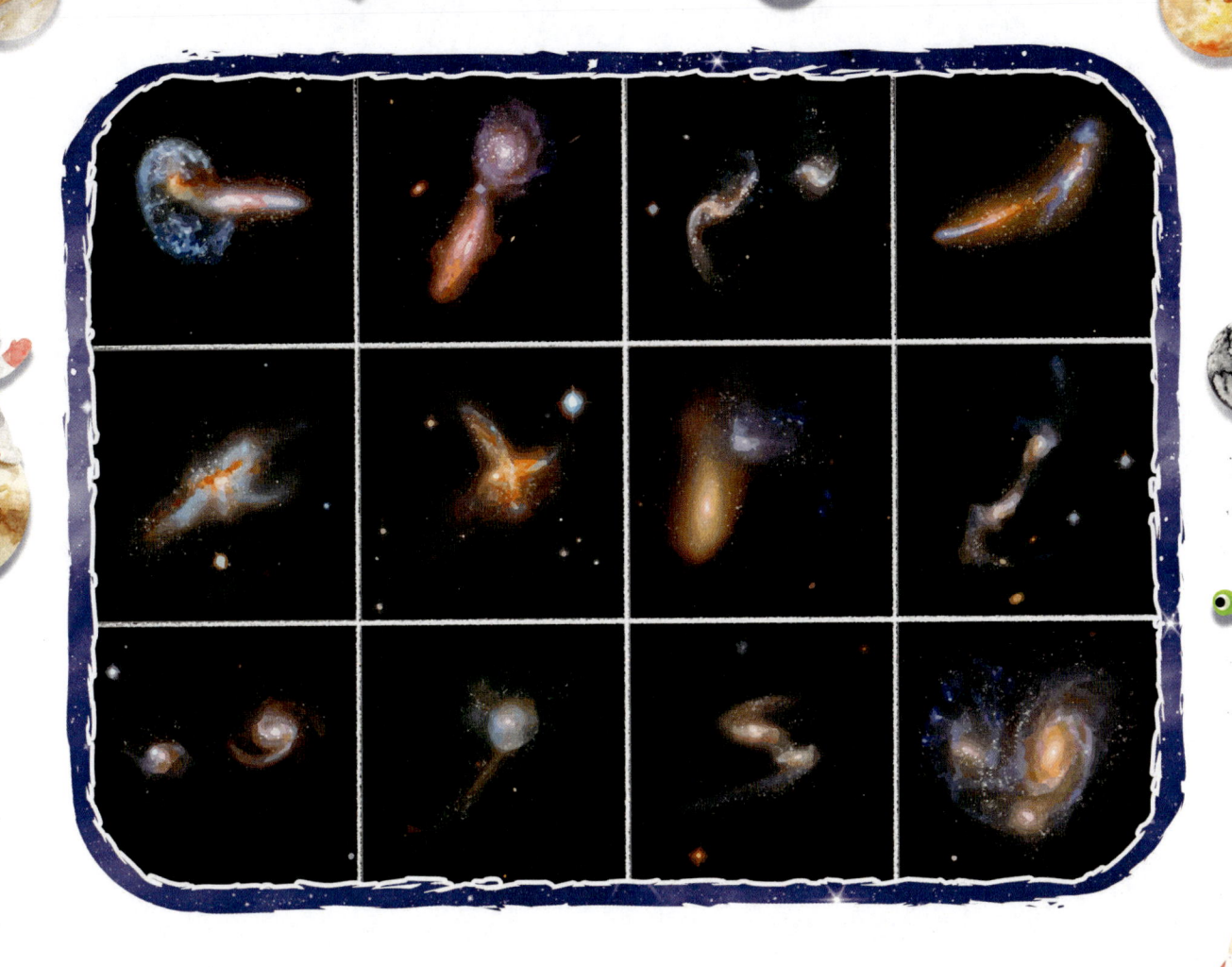

GALAXIES

The dictionary describes galaxies as a system of millions or billions of stars, together with gas and dust, held together by gravity. You may have heard people describe galaxies as a "star system," "solar system," "constellation," "cluster," "nebula," "spiral-galaxy," "Seyfert galaxy," "stars," or "heavens."

There are many galaxies in the universe there is one called the "spiral galaxy." One of the most popular galaxies is the "black eye galaxy." Scientist estimate that there are over one trillion galaxies. Planet earth lives in the Milky Way galaxy.

Some galaxies have some of the craziest planets too. Many galaxies have super earths, and planets that rain rocks. There are too many galaxies to count. Scientist only know a couple compared to how many there are in space.

Activity #1: Research, illustrate, and label a galaxy of your choice below.

PLANETS

Did you kn ow that Mars has the biggest volcano we know of? Some planets are considered to be "gas planets." Saturn, Jupiter, Neptune and Uranus are examples of gas planets.

There are planets that orbit their own star; and have their own moons. For example, Jupiter has a moon that scientist have named, "Ganymede."

Planets like Saturn and Neptune rain diamonds. Other planets rain rocks. Some planets rain rocks instead of water.

because it gets so hot. So hot in fact, that some rocks evaporate before hitting the surface.

Planets have funny names. One of them was given the name, "Ogle tr 56b" and it rains iron. Many other ones have methane rain. Sulfuric acid rain was discovered on Venus. These are only a few examples of how different the weather and rain is on other planets. Cool huh!

Activity #2: iLLUSTRATE and LABEL the solar system.

THE SUN

The sun is located at the center of the Solar System. According to Quizlet.com, the Sun is a "nearly perfect sphere of hot plasma, with internal convective motion that generates a magnetic field through a dynamo process." Those big words mean the sun is hot and full of energy that powers the whole solar system. Scientists believe it is the most important source of energy for life on Earth.

The sun's rotation time at the equator is about 27 days, and the rotation time at "the poles" (The North and South Pole) is about 36 days.

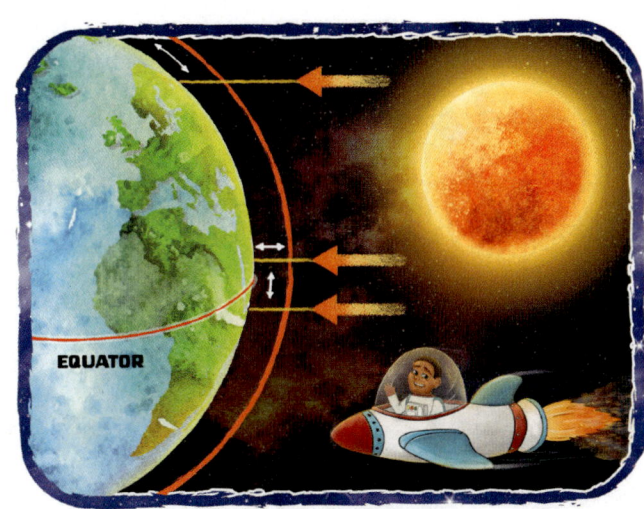

11

THE STRUCTURE OF THE SUN

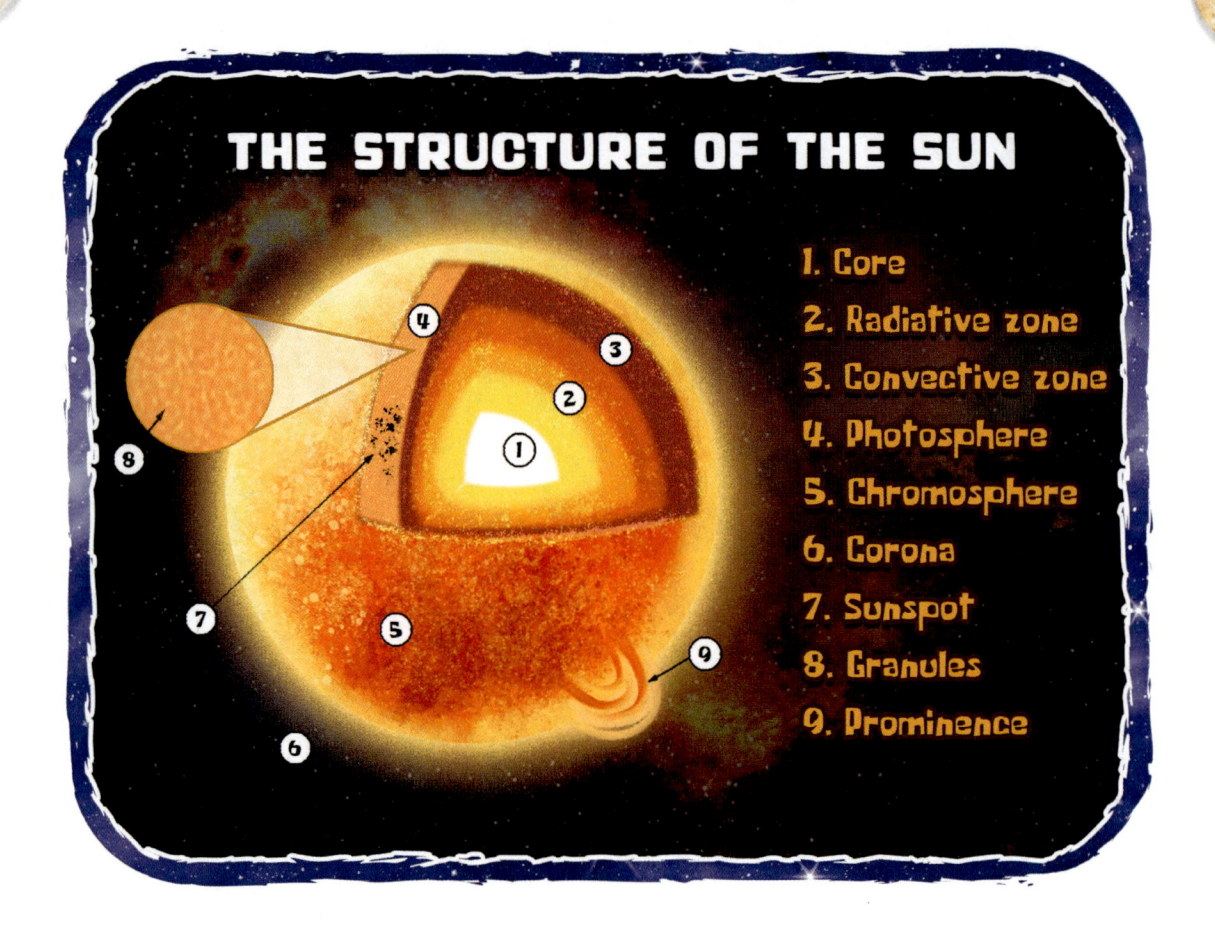

1. Core
2. Radiative zone
3. Convective zone
4. Photosphere
5. Chromosphere
6. Corona
7. Sunspot
8. Granules
9. Prominence

Some people, my dad included, thinks the sun is a planet. SHOCKER! it's not. The sun is made of mostly hydrogen and helium. it is a ball of gas and its main category is a star and has nine layers to its structure.

Some people would assume that the sun is the hottest star in space but wrong again! The hottest (not necessarily largest) star known to scientists so far, is the Blue star. The second hottest burning star is the Red star. The sun is in third place as the hottest star in space though one of smallest when compared to others.

OUR SUN IS THE MERE PIXEL YOU CANNOT SEE BESIDE ALBIREO THE DOUBLE STAR

SUN
ALBIREO
KOCHAB
RIGEL
DENAB
THE PISTOL STAR
RUBY STAR
ANTARES

My teachers say things like, "shoot for the stars!" i'm not sure i want to touch one now! Too hot up there and they live forever!

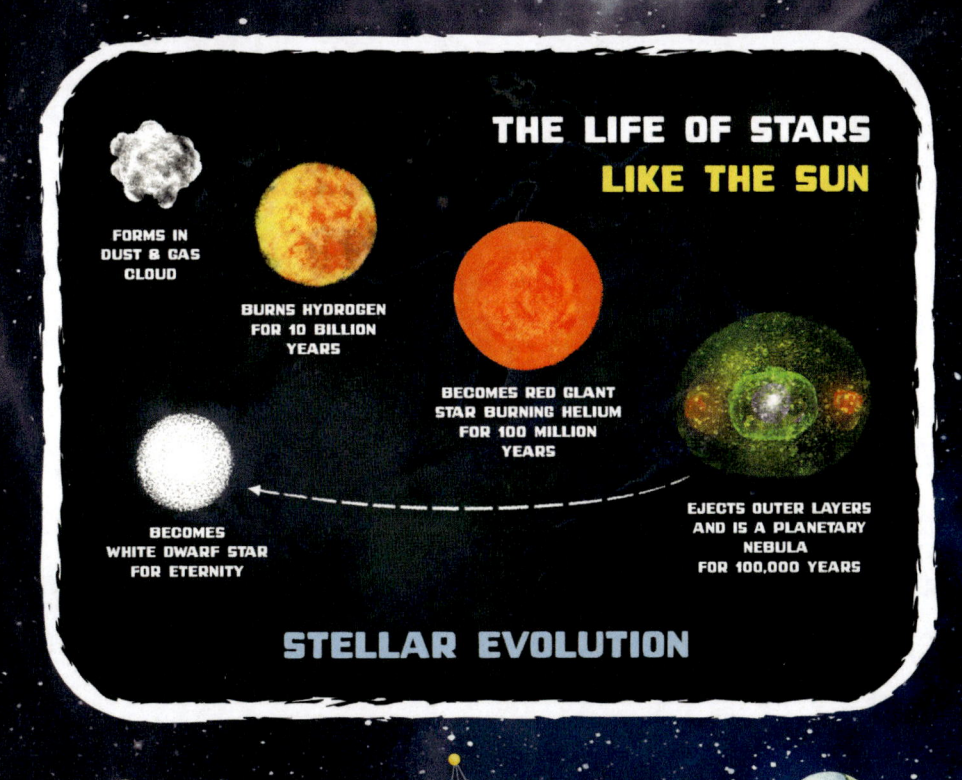

THE LIFE OF STARS LIKE THE SUN

- FORMS IN DUST & GAS CLOUD
- BURNS HYDROGEN FOR 10 BILLION YEARS
- BECOMES RED GIANT STAR BURNING HELIUM FOR 100 MILLION YEARS
- EJECTS OUTER LAYERS AND IS A PLANETARY NEBULA FOR 100,000 YEARS
- BECOMES WHITE DWARF STAR FOR ETERNITY

STELLAR EVOLUTION

BLACK HOLES

A black hole is created when a star dies. Black holes' secret weapon is gravity. Scientists have said black holes exist 250 million light-years from earth. Black holes remind me of tornadoes. i say tornadoes because of its gravitational pull, sucking you into it like dirt in a vacuum cleaner. it can gobble objects up, and throw anyone, or anything, anywhere.

SOLAR FLARE

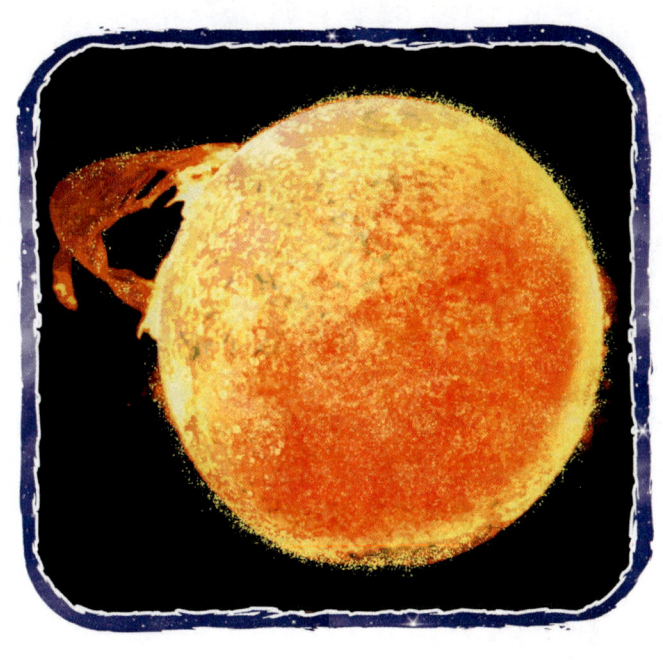

A solar flare is a dangerous event in space. A solar flare is a sudden flash of brightness observed near the Sun's surface. it's the sun erupting hot energy like a volcano, throwing or spitting fire and heat in to space. if the flare should hit near earth's atmosphere it could cause damage to satellites. if satellites get damaged, it could mean BYE-BYE WiFi connection!

if solar flares hit planet earth, it can cause forest fires, droughts, an extreme lack of water and leaving many people with NO FOOD!

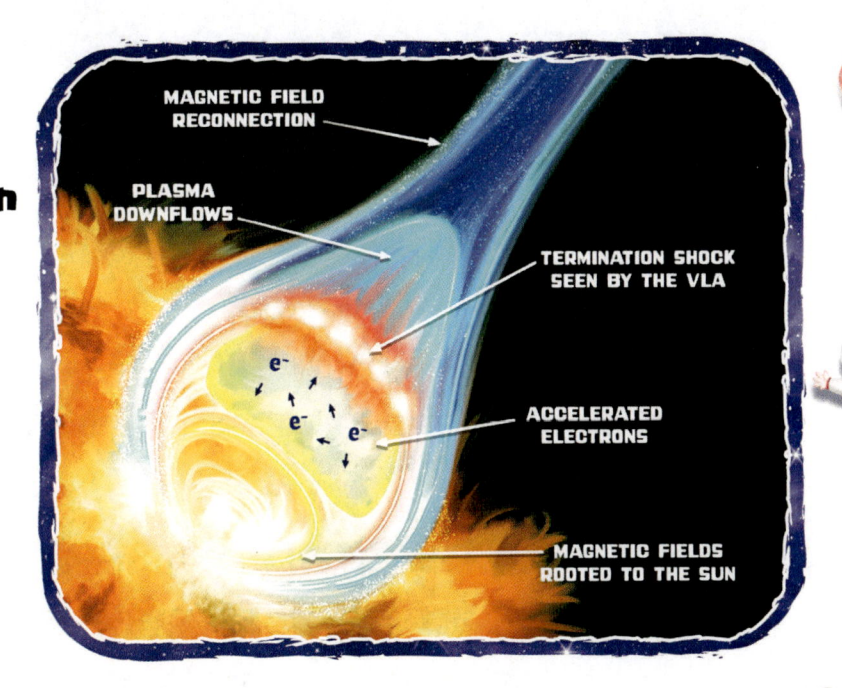

MAGNETIC FIELD RECONNECTION
PLASMA DOWNFLOWS
TERMINATION SHOCK SEEN BY THE VLA
ACCELERATED ELECTRONS
MAGNETIC FIELDS ROOTED TO THE SUN

SPACE DEBRIS

Space debris are pieces of rock and trash from different planets and satellite parts. it could also include space ice or hardware from broken satellites. Another name for debris is space waste or junk. When large amounts of space debris stick together, they can form large rocks called asteroids. Some asteroids get so large, they can form their own moons from the debris.

Meteors and comets can also be formed by space debris traveling at high speeds. Space debris can orbit and even form rings around a planet. For example, Saturn's rings are made of space debris. You will also see a lot of comets near there.

As you can see close up, Saturn's rings have space debris that orbits around the planet.

METEORS, ASTEROIDS & COMETS
"THE SHOOTING STARS"

When people think of meteors, they think of huge crater tsunamis crashing into earth destroying people, their homes and more. This is because we typically see this in movies and television.

Movies sometimes show meteors catching on fire from the earth's atmosphere. This is true. The most common scenes are meteors causing major damage to the earth which isn't common at all. Sometimes special effects in movies have meteors sometimes glowing, or electricity surrounding it. This is not far from the truth. Meteors and meteorites actually store electric molecules.

What many do not know is in reality, meteors are not that dangerous to planet earth. They are the smallest compared to comets and asteroids. Second, there is a big difference between "meteoroids" and "meteorites." For example, meteoroids are small rocks in outer space. if meteoroids make it through the earth's atmosphere, they would be called "meteorites" instead of meteoroids. The only difference is the location that determines their name changing.

Meteoroids travel at high speeds. When entering the earth's atmosphere it immediately starts to burn the rock. The reason for the streak of light seen at night is the rock burning from the earth's atmosphere, looking like "shooting stars."

The earth's atmosphere melts most of the meteor before impact. There is nothing to fear. However, some survive the journey and land on earth.

If they did hit earth, they could create huge craters. If meteorites hit by sea, it can create huge tsunamis. Or worse, meteorites can cause a nuclear explosion!

SO RUN IF YOU SEE THESE COMING!

There are major differences between asteroids, meteors and comets. Asteroids orbit planets like Mars and Jupiter along what scientist call, the "asteroid belt." On the other hand, comets like asteroids, also orbit the sun though in a different pattern. Meteors not so much. Lastly, asteroids are large masses made of rock, or metal much larger than a meteor.

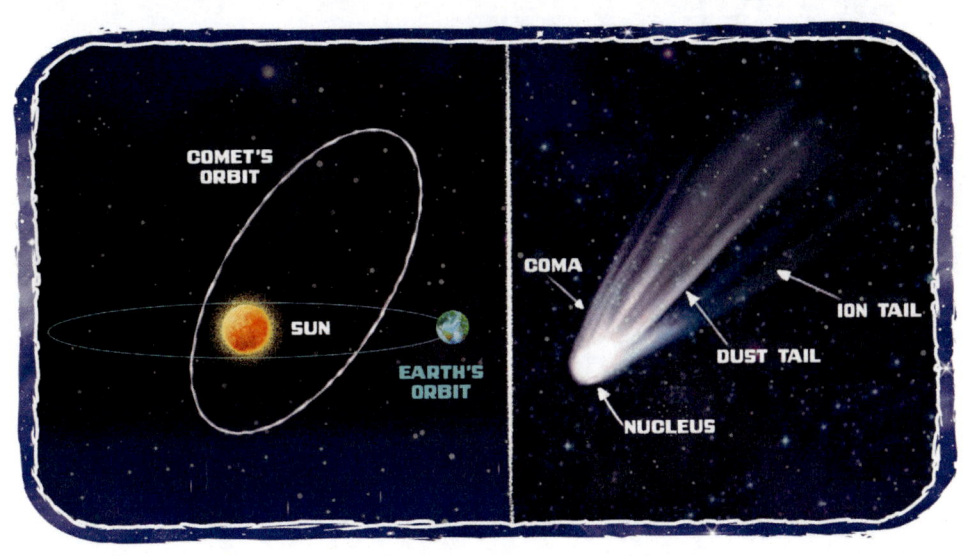

Really quick about comets. They have a tail, made from space dust, ice, debris and gas -not rock. Dust from debris, ice from water, carbon dioxide, ammonia, methane and gases make comets different from the large rock and metal masses of asteroids. Meteors can be formed from both comets and asteroids.

SUPERNOVA

A supernova is when a star explodes and expands over several weeks and possibly months. There are three main things that happen. The explosions are so wonderful and big, scientists have confused them with galaxies.

The process is very cool!

1. First, the star dies, or explodes.
2. Then, the star forms into a supernova.
3. Last, the supernova becomes a black hole.

However, after the nova, do not go near the location unless you want a deadly surprise.

SCIENTIFIC QUETION: i wonder how supernovas impact humans, planet earth, and earth's atmosphere?

SOLAR ECLIPSE

A solar eclipse occurs when the moon gets between Earth and the sun. The moon casts a shadow over Earth, blocking the rays from the sun.

A solar eclipse can only take place at the phase of the new moon.

A solar eclipse is not dangerous but if you stare at it long enough, you can go blind.

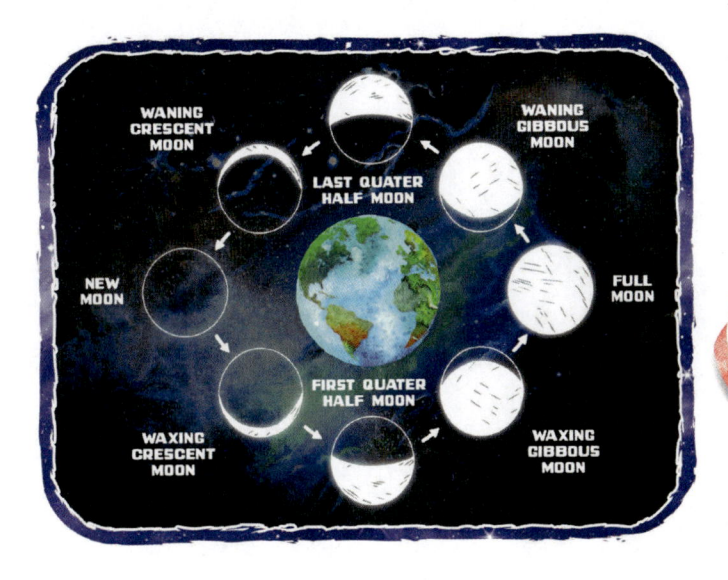

it only happens once a century (a century = 100 years) so better get to watching when there is one.

JOSIAH'S QUIZ

(Use complete sentences)

1. List three synonyms for galaxies.

2. Saturn's rings are made up of what?

3. Which planets are considered "gas" planets?

4. Compare and contrast asteroids, meteors and comets.

5. What happens when stars become supernovas?

6. What part of the moon phases can a solar eclipse form? And when does a solar eclipse occur?

7. Describe and illustrate a star's life cycle.

8. Do comet's have their own orbit around planets, the sun, or both? Explain the orbit differences compared to other objects?

9. What is it called when the sun spits fire in space? How could it impact earth?

10. What is the "secret weapon" of black holes?

Glossary

Constellation - A constellation is a group of stars that forms an imaginary outline or pattern on the celestial sphere, typically representing an animal, mythological person or creature, a god, or an inanimate object. The origins of the earliest constellations likely go back to prehistory.

Convection - is the transfer of internal energy into or out of an object by the physical movement of a surrounding fluid that transfers the internal energy along with its mass.

Dynamo process - The dynamo theory describes the process through which a rotating, convecting, and electrically conducting fluid can maintain a magnetic field over astronomical time scales. A dynamo is thought to be the source of the Earth's magnetic field and the magnetic fields of other planets.

Equator - An equator is the intersection of the surface of a rotating sphere with the plane that is perpendicular to the sphere's axis of rotation and midway between its poles. The equator usually refers to the Earth's equator: an imaginary line on the Earth's surface equidistant from the North Pole and South Pole, dividing the Earth into the Northern Hemisphere and Southern Hemisphere.

Galaxy - A galaxy is a gravitationally bound system of stars, stellar remnants, interstellar gas, dust, and dark matter. The word galaxy is derived from the Greek galaxias, literally "milky", a reference to the Milky Way. Galaxies range in size from dwarfs with just a few hundred million stars to giants with one hundred trillion stars, each orbiting its galaxy's center of mass

Magnetic field - A magnetic field is the magnetic effect of electric currents and magnetic materials.

perseus - Perseus is a constellation in the northern sky, being named after the Greek mythological hero Perseus. it is one of the 48 ancient constellations listed by the 2nd-century astronomer Ptolemy.

Solar Flare - A brief eruption of intense high-energy radiation from the sun's surface, associated with sunspots and causing electromagnetic disturbances on the earth, as with radio frequency communications and power line transmissions

Space ice - interstellar ice consists of grains of volatiles in the ice phase that form in the interstellar medium. ice and dust grains form the primary material out of which the Solar System was formed. Grains of ice are found in the dense regions of molecular clouds, where new stars are formed.

Super Earth - A super-Earth is an extrasolar planet with a mass higher than Earth's, but substantially below the mass of the Solar System's ice giants Uranus and Neptune, which are 15 and 17 Earth masses respectively. The term super-Earth refers only to the mass of the planet, and does not imply anything about the surface conditions or habitability.

Tsunamis - A tsunami or tidal wave, also known as a seismic sea wave, is a series of waves in a water body caused by the displacement of a large volume of water, generally in an ocean or a large lake. Earthquakes, volcanic eruptions and other underwater explosions above or below water all have the potential to generate a tsunami. Unlike normal ocean waves, which are generated by wind, or tides, which are generated by the gravitational pull of the Moon and the Sun, a tsunami is generated by the displacement of water.

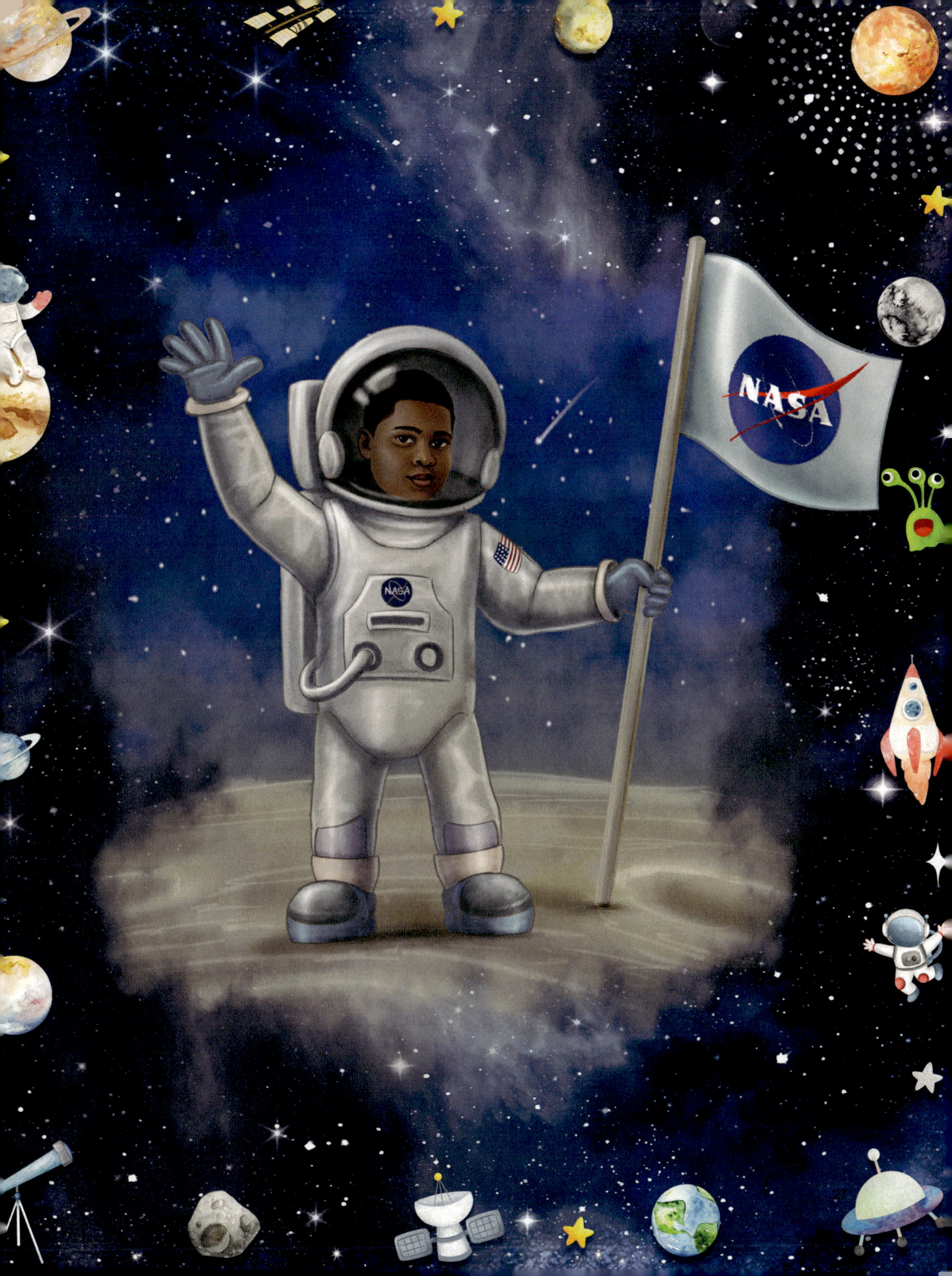

SPECIAL PROJECTS: BEYOND THE TEXT
Essay Assignment

Prompt A: Choose one of the concepts from the text and explain its impact on Earth's atmosphere, surface and human beings.

Prompt B: Compare and contrast at least one other planet to Earth. Categories such as weather, resources, life, terrain and physical traits.

Instructions:

- 5 paragraphs = introduction, 3 supporting paragraphs and conclusion
  - 12 Font typed in Times New Roman double-spaced
  - 3 expert sources (this book does not meet expert criteria) An expert has worked at least 5 years in their profession
  - Paragraphs = At least three sentences
    - A topical sentence.
    - A sentence with specific knowledge or evidence.
    - A sentence explaining what the evidence proves and/or how the audience should interpret the evidence

introduction paragraph must have a thesis statement that contains three parts: a topic, claim and 3 major points. The 3 major points in thesis statement will be the focus in the supporting paragraphs.

Supporting paragraphs should have at least three sentences

o Topical sentence explaining the topic in that paragraph (the topic is one of your "major points" in your thesis statement)

o Evidence using a quote from an expert to support your thesis statement. After the last quotation mark closing the quote, place in parentheses a citation with the author's last name, a comma, and year of publication after the quote.
Example: (King, 2019)

o The last sentence should explain what the evidence/quote you used proves in regards to your claim, noted in the introduction

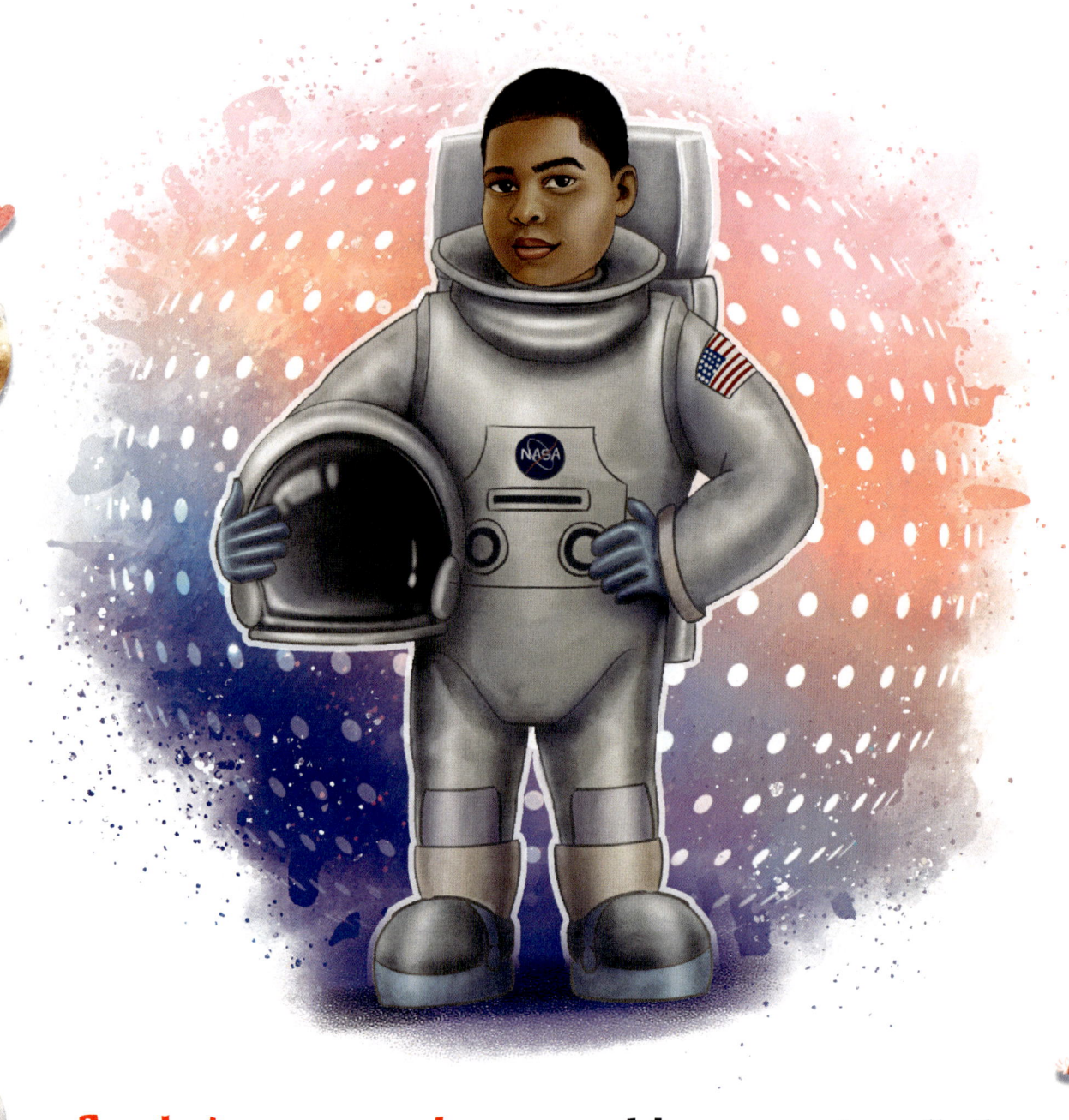

Conclusion paragraphs are used to summarize all of your major points in supporting paragraphs 1-3. Also, its an opportunity to give perspective on what could and should happen leading to better outcomes.

Create a 3D Model

Create a three-dimensional model by choosing a planet to study.

The model must be labeled (life, physical traits, weather samples, planet nicknames, resources, terrain, outer rings, moons etc.) along with other useful facts for the audience.

Other options for three dimensional models include any of the concepts expressed in this book.
For example:

- Supernova
- Solar Flares
- Black Holes
- Galaxies
- Solar Eclipse
- The Sun
- Meteors
- Asteroids
- Comets
- Space Debris

SOURCES for KIDS

- www.britannica.com
- www.dictionary.com
- www.nasa.gov

EXTRA CREDIT

MR. KING'S TUTORIAL TEST KEY

Please look at question #1 for guidance on how to answer questions in complete sentences. i use a research writing process that varies in grade level. Students from grades three to eight can use this method for school, essay exams, or improving their formal writing skills. Gradually this test key helps to develop CRITICAL THINKING SKILLS.

You will notice that there are only four sentences but, they are very specific. it is not about quantity as much as it is about quality of work.

Each answer contains FIVE components:

- A claim / thesis / educated opinion from the required reading
- Location of evidence used to answer the question
- Evidence in the form of a quote from the text that answers the question
- A conclusion statement summarizing your evidence
- A citation to give credit to the author's work and ideas expressed in theirbook.

1. (WRITE A CLAIM / THESIS / EDUCATED OPINION FROM READING) Galaxies are a large group of stars commonly described with many names. (WHERE YOU FOUND YOUR ANSWER) My evidence comes from paragraph one, on page nine. (PROVIDING A QUOTE FROM THE TEXT AS YOUR EVIDENCE) King writes, "You may have heard people describe galaxies as a "star system," "solar system," "constellation," "cluster," "nebula," "spiral galaxy," "Seyfert galaxy," "stars," or "heavens." (CONCLUSION) We learn from King that galaxies have many names that describe it.(CITE YOUR SOURCE) (King, 2019).

2. Saturn's rings are made of many different things like space rocks, trash from satellites, and ice. My evidence comes from paragraph ___, on page ___. According to King, "Space debris can orbit and even form rings around a planet. For example, Saturn's rings are made of space debris." King informs his audience that other planets have pollution with space trash from satellites, mixed in with the nature, just like planet earth.(_____, 2019).

3. All planets are not gas planets but there are some that have an unusual amount of gases. My evidence comes from page ____, paragraph ___. The author states, "Some planets are considered to be "gas planets." Saturn, Jupiter, Neptune and Uranus are examples of gas planets." We learn from the author that planets are not only large, but have different characteristics. One of which would be the amount of gas from planet to planet.(King, ____).

4. ADVANCED LEVEL - Compare and contrast answers are typically longer, require more detail and multiple paragraphs even with short answers because of the length of the information involved. (ANSWER IS BELOW)

      Meteors, asteroids and comets are often confused because they have similar functions, stay close together and in some cases, form from each other's physical properties. When studying them closely, they are distinct from one another in many ways such as orbit and physical make-up. My evidence comes from page ____ and ____. King says, "Asteroids orbit planets like Mars and Jupiter along what scientist call, the "asteroid belt." On the other hand, comets like asteroids, also orbit the sun though in a different pattern. Meteors not so much." What we learn from King is asteroids and comets do orbit though not is the same way. Meteors just float as pieces from asteroids and perhaps comets colliding.

      Physically all three of them are different. Asteroids are large, comets are large, but meteors are not. Their shapes range in variety too. My evidence comes from page _____, last paragraph. According to King, "[Comets] have a tail, made from space dust, ice, debris and gas -not rock like asteroids. Dust from debris, ice from water, carbon dioxide, ammonia, methane and gases make comets different from the large rock and metal masses of asteroids. Meteors can be formed from both comets and asteroids components." (2019).

      King reminds his audience to consider taking a closer look at these three things in outer space scientifically as they are, and not in the generalized manner we see them from television or movies.

5. Evidence = Found on page ____ - "First, the star dies, or explodes. Then, the star forms into a supernova. Last, the supernova becomes a black hole."

6. Evidence = Found on page ____ - "New Moon Phase" / "A solar eclipse occurs when the moon gets between Earth and the sun."

7. Evidence = See graph on page _____.

8. Evidence = See section on meteors, comets and asteroids on page ___ and ____. Look at the illustrations of orbit to describe the specific patterns of comets.

9. Evidence = See section on Solar Flare (page ____)

10. Evidence = See section on Black Holes (page ____)

NOTES

King's Press Publishing

Real people. Powerful stories. Infinite Possibilities

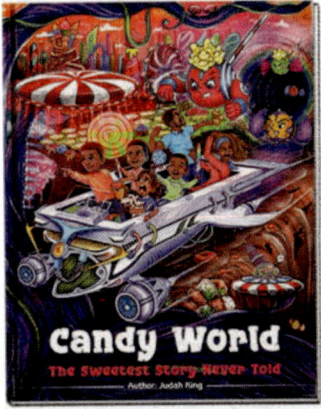

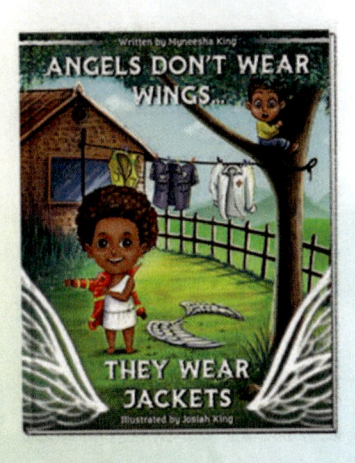

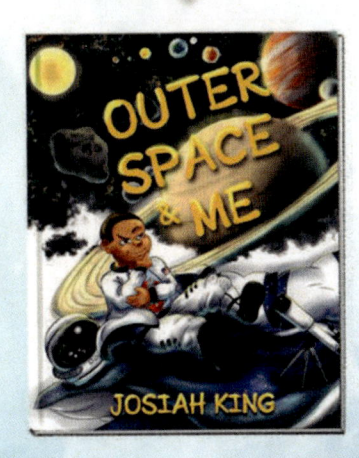

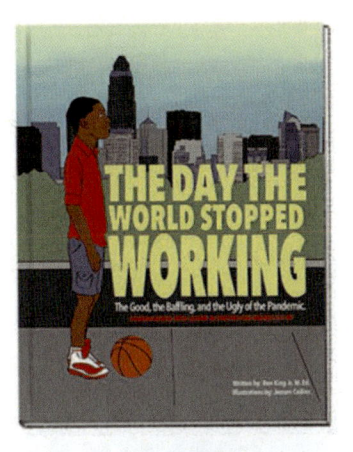

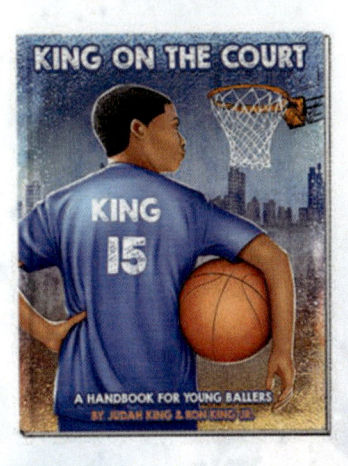

www.kingspresspublishing.com

Made in the USA
Middletown, DE
30 October 2022